ISBN 978-1-5278-5586-1
PIBN 10891181

1 MONTH OF
FREE
READING

at

www.ForgottenBooks.com

By purchasing this book you are
eligible for one month membership to
ForgottenBooks.com, giving you
unlimited access to our entire
collection of over 1,000,000 titles via
our web site and mobile apps.

To claim your free month visit:
www.forgottenbooks.com/free891181

English
Français
Deutsche
Italiano
Español
Português

www.forgottenbooks.com

Mythology Photography **Fiction**
Fishing Christianity **Art** Cooking
Essays Buddhism Freemasonry
Medicine **Biology** Music **Ancient**
Egypt Evolution Carpentry Physics
Dance Geology **Mathematics** Fitness
Shakespeare **Folklore** Yoga Marketing
Confidence Immortality Biographies
Poetry **Psychology** Witchcraft
Electronics Chemistry History **Law**
Accounting **Philosophy** Anthropology
Alchemy Drama Quantum Mechanics
Atheism Sexual Health **Ancient History**
Entrepreneurship Languages Sport
Paleontology Needlework Islam
Metaphysics Investment Archaeology
Parenting Statistics Criminology
Motivational

Creosoting

. . . Timbers

By WM. B. McKENZIE,

Member Canadian Society C. E., Member American
Society C. E., Assistant Engineer
Intercolonial Railway.

Reprinted from CANADIAN ENGINEER.

CREOSOTING TIMBERS.

BY WILLIAM D M'KENZIE, MEM. CAN. SOC. C.E., MEM. AM. SOC. C.E., ASST. ENG. INTERCOLONIAL RAILWAY.

Timber.—It is of the utmost importance that only the species of timber best adapted for receiving the creosote should be used. Of the thirty-five different kinds of pine found in the United States and the ten kinds which grow in Canada, besides the Douglas fir or Oregon pine, as far as known at present, only the short leaf *Pinus mitis*, Michx., and the Loblolly-pine, *Pinus toeda*, Linn., are suited for creosoting as a protection against marine insects. The Short-leaf pine is found in great perfection in the light sandy soil of Virginia and the northern part of North Carolina, while the Loblolly flourishes on the lower ground near the coast of both States. All the other pines, as well as the Douglas fir (Oregon-pine), spruce and hemlock, have too little sap wood for successful creosoting. They are also variable in texture, and require such a high and long-continued heat, that the wood is checked in the cylinders and the fibre injured. These two species of pine are known by several different local or common names in different places, for instance : *Pinus mitis,*

Michx., or *Pinus echinata*, Miller, commonly known as Short-leaf pine, North Carolina-pine, Yellow-pine, Spruce-pine, Bull-pine, Rosemary-pine, Frankincense-pine, Sweet-pine, Sap-pine, Loblolly-pine, Oldfield-pine, Slash-pine. Color, yellowish red. Sapwood, commonly over four inches of the radius. Section variable. Rings wide near the heart, followed by zone of narrow rings, not less than four, mostly eleven or twelve rings to the inch. A moderate amount of resin. This tree prefers a well-drained, light, sandy or gravelly soil, or warm light loam. Foliage short and scant. Cones small. Bark reddish, in long plates. Crown pyramid-shaped.

Specific gravity	0.6104	
Percentage of ash	0.2900	
Weight per cubic foot (lbs.)	38 04	at 212° Fah.
Compressive strength with grain	5,900	lbs. per sq. inch.
Compressive strength across grain	940	" "
Bending strength	9,230	" "
Tensile strength	13,400	" "
Shearing strength	688	" "

Pinus toeda (Linn.) commonly known as Loblolly-pine, Virginia-pine, Short-leaf pine, Rosemary-pine, Frankincense-pine, Indian-pine, Oldfield pine, Bastard-pine, Slash-pine, Black-pine, Swamp-pine, Meadow-pine, Sap-pine, Cornstalk-pine, Foxtail-pine. Height of mature trees, 125 to 150 feet. Foliage thin. Color of foliage, sea-green. Coarse grain, 3 to 12 rings per inch, generally wider than in *Pinus mitis*. Color whitish to brownish yellow, the dark bands of summer wood being proportionately narrow. Sap wood variable, one-third to one-half of the radius. Resin abundant about midway between short-leaf and long-leaf. Bark grayish, in deeply fissured plates.

Specific gravity	0.6343		
Weight per cubic foot (lbs.)......	39.23	at 212ᵛ Fah.	
Compressive strength with grain..	6,500	lbs. per sq. inch.	
Compressive strength across grain	990	"	"
Bending strength...'..............	10,100	"	"
Tensile strength	14,100	"	"
Shearing strength	690	"	"

GENERAL INFORMATION ON PINES.

The annual rings are closer together at the top of the tree. Logs cut from the foot of the tree are 7 per cent. stronger, and two pounds per cubic foot heavier. The greater the weight, the greater the strength. The strongest wood is at one-third the distance from the heart, and the strength decreases from the heart to the periphery 15 per cent. to 25 per cent. Large beams are from 10 per cent. to 40 per cent. weaker than small beams of the same material. Green timber beams fail first on the compression side. Seasoned wood is 50 per cent. to 100 per cent. stronger than green wood. Short-leaf pine is one-third weaker than long-leaf pine. Wood seasoned out of doors under shelter, retains about 15 per cent. of moisture, computed on the dry weight. Wood used in doors retains about 10 per cent. of its moisture, computed on the dry weight. The faster the drying, the greater the checking and warping. Boxed or tapped long-leaf pine timber is slightly stronger than untapped timber, and is in no respect inferior. The tapping extends over a period of four years. A large proportion of the long-leaf pine lumber is from tapped trees, and it is never kept apart or distinguished from the untapped by either the millers or the dealers. No available criteria exist by which to distinguish the two kinds of long-leaf timber—tapped and untapped—

after manufacture. Some of the most resinous long-leaf timber comes from tapped forests and some of the driest from untapped forests. Age or use does not destroy the strength of timber, unless decay or season-checking takes place. The heaviest wood shrinks most in drying. Top-logs shrink 15 per cent. to 20 per cent. less than butt-logs. Short-leaf pine shrinks one-quarter less than long-leaf pine. The amount of water in the wood does not rule the shrinkage. Wood is composed of carbon, 52.4 parts ; hydrogen 5.7 parts, and oxygen, 41.9 parts, and its specific gravity is about 1.5. The bark forms 10 per cent. to 15 per cent. of the volume. It is only owing to the looseness of texture that most timbers are lighter than water. Immersion in fresh water soon after felling makes timber more durable, and one year in salt water doubles the life of timber. Sap freshly drawn from a tree begins to putrefy in 24 hours. The time of felling has no effect upon the strength of pines. Green sap wood contains about 50 per cent. water computed on the dry weight. Sap wood is heavier than heart wood, and the part formed in the summer is twice as heavy as that formed in the spring. Sap wood contains 1 per cent. to 4 per cent. of resin and about 1-6th of the resin is composed of turpentine. It is from the sap wood of the long-leaf pine only that resin is obtained by tapping. Sap wood shrinks one-quarter more than heart wood. In trees 100 years old sap wood changes to heart wood in 30 to 60 years. In trees 25 years old sap wood changes to heart wood in 70 to 80 years. In trees between the age of 80 and 100 years the sap wood is the strongest. The sap wood in short-leaf and Loblolly pines forms about 60 per cent. of the volume. Unseasoned heart wood contains about 20 per cent. water computed on the dry weight. Heart wood of long-leaf pine contains 5 per cent. to 24 per cent. of resin, and about 1-6th of it is composed of turpentine. This resin is thick, and will not flow when the tree is tapped. The cells of the heart wood are dead and contain only water or air. At the top of the tree the heart wood is the strongest. Excessive steaming or heat over 250° Fah., destroys the elasticity of timber. Creosoted timber is less liable to burn than untreated timber. Creosoting with dead oil of coal-tar increases the resistance to bending 15 per cent., and the resistance to compression 22 per cent. Piles driven after the 15th June are seldom attacked by the *teredo* until the following spring.

PREPARATION OF PILES.

The sap is more fluid in timber cut in the first months of the year, and if the piles are floated direct from the forest to the creosoting works, water will enter the cells, drive out part of the air and act as a solvent to liquefy the gummy parts of the wood; so that when they are hauled out of the water, barked and laid on skids to dry, they will season better and more quickly than piles treated in any other way.

STEAMING.

The creosoting cylinders are usually 6 feet diameter by 100 feet long, of $\frac{5}{8}$ inch steel, fitted with cast-iron heads weighing three tons each, securely bolted on the ends of the cylinders. When the piles for treatment have been run into these cylinders on trolley cars, the doors are closed and bolted. Steam is then admitted to the cylinders and its heat increased by steam-heated pipes in the bottom. When the temperature in the wood reaches 187° Fah., the albumen in the sap (about one per cent.) solidifies. The heat enters slowly from the surface towards the interior, and the sap is made more fluid. The air expands, and part leaves the cells, and, as the temperature rises above 212° Fah., the water therein is gradually transformed into steam. The extractive ingredients of the sap are driven from the wood, and, together with the water of condensation, collects at the bottom of the cylinder. The time required is directly proportional to the diameter and density of the timber, and varies from 8 to 10 hours. The temperature of the steam should not be allowed to rise over 250° Fah., as higher temperatures or long-continued steaming softens and separates the fibres of the wood, and thus reduces its strength.

VACUUM.

After the steam has been on a sufficient length of time, say fom 8 to 10 hours, it is allowed to flow out of the cylinders by its own pressure, and this is followed by the vacuum pump exhausting the air and vapor from the cylinders and the wood, removing at the same time the extractive ingredients of the sap from the cylinders. A vacuum of from 22 to 25 inches is continued for from three to five hours, and until the discharge has neither odor nor taste of turpentine. In order that as much as possible of the water in the cells towards the centre of the timber be volatilized and removed as steam by the vacuum pump, it is necessary that while the pump is working the heat be kept above the condensing point, which varies from a temperature of 212° F. under one atmosphere, down to 135° F. under a 25-inch vacuum. The heat must not be so high as to cause checking of the timber. Short-leaf pine withstands this ordeal better than any other kind of wood. While the vacuum is on, and when the timber is green, steam is usually admitted to the coils within the cylinders at about 250° to 300° F., and when the wood is very wet, at a maximum of 350° F.

Filling the Cylinders with Oil.—While the vacuum is still on, the creosote heated by steam-pipes to about 125° Fah., is allowed to run by gravitation from the gauge tank, until the cylinders are completely filled. The creosote should be maintained at a temperature of about 125° Fah. during the treatment. It is important that no water from condensed steam or otherwise should be allowed to mix with the creosote as it rises to the top of the cylinders, and in this case the wood at the top only receives an injection of dirty water. The cylinders will run about 3-5ths full by gravitation; then, with the aid of the force-pump, they are completely filled with creosote. The valves are now closed and the reading of the gauge-tank noted; the pressure is applied by the force-pump and the creosote is forced into the empty cells of the wood under a pressure of from 110 to 135 lbs. per square inch, which pressure is kept up until the specified number of pounds of oil has been forced into the timber as indicated by the gauge-tank; the usual time required being from three to six hours for piles of short-leaf or Loblolly pine. The pressure is then released, the unabsorbed creosote in the cylinders forced by the air-pump back into the gauge-tank, the cylinder doors opened and the timber removed. From the taking out of one charge of 10 to 20 piles to the putting in of another, the time consumed is usually about 20 to 27 hours. The quantity of oil absorbed by the timber is measured by determining the difference in volume of the oil in the gauge-tank before and after the operation, and as $2\frac{1}{4}$ or $2\frac{1}{2}$ lbs. of creosote is absorbed per cubic foot of timber, while the cylinders are filling, this amount should be deducted from thē specified quantity.

Creosote.—The production of the dead oil of coal tar in the United States is insufficient for the needs of the country, and a considerable quantity is imported from England. What is commercially known as "London oil," a thick and heavy oil, is considered to be the best produced in England for marine work. Creosote at 65° Fah., weighs about 8 to 9 lbs. per U. S. gallon, and boils from 380° to 760° Fah. Analy· sis of American and English oils used in 1895 for treating piles are as follows :

AMERICAN OIL.

The sample as received, well mixed, contains water   0.18 per cent.
Oils (lighter than water distilling over between 350° and 410° Fah., carbolic acid, creosote, etc.) 1.13
Oils (heavier than water distilling between 410° and 540° Fah., naphthaline, etc., crystalline.. 73.10   "
Higher phenoloid bodies, distilling between 540° and 610° Fah.. .... 14.67
Heavy crystalline substance and a little red oil distilling between 610° and 680° Fah........ ·6.79
Soft pitch, not volatile at 680° Fah............ 4.13

                            ———
                            100.00

ENGLISH OIL.

The sample, as received, well mixed, contains water... 0 25 per cent.
Oils (lighter than water, distilling over between 392° and 450° Fah., phenol)............... 9.50
Oils (heavier than water, distilling over between 450° and 482° Fah., naphthaline, crystalline when cold).. .... 24.00 per cent.

Oils (heavier than water, distilling over between 482° and 540° Fah., naphthaline, crystalline when cold)..	28.50 per cent.
Higher phenoloid bodies, distilling over between 540° and 610° Fah......................	10.25
Heavy crystalline substance and a little red oil, distilling over between 610° and 680° Fah...	12.00 "
Soft pitch, non-volatile at 680° Fah..............	15.50 "
	100.00 "

No two lots of oil will give precisely the same analysis, so that only general qualities should be called for in specifications. The phenols, which include crude carbolic acid, cresylic acid, and other tar-acids, are the germ-destroyers, and some of the heavier constituents, principally naphthaline and acridine, which crystallize in the cells and render the wood water and air-proof, are the germ-excluders.

About 10 lbs. per cubic foot is sufficient to prevent decay above water, and 15 or 16 lbs. per cubic foot to protect the piles against the ravages of marine insects for at least thirty and perhaps fifty years, in Canadian waters. Creosoted piles at Sydney, Cape Breton, in use 24 years, are still in perfect condition.

The following quantities of creosote per cubic foot are considered a sufficient protection against sea-worms at the places mentioned below :

In English harbors..	10 to 12 lbs. per cubic foot.
Northern harbors in the United States	10 to 12 "
Holland and Belgium...............	10 to 12
France........................	19
Gulf of Mexico............	20
Canada......................	15 to 16 "

The square timber should be handled carefully after creosoting, to avoid chafing or brooming the edges and thus spoiling its appearance.

Economy in creosoting.—Piles can be treated most economically at a shipping port near the forest in which they grow, because the following items of loss are thereby avoided :

1. The handling and the freight on the bark, which is about 10 per cent. or 15 per cent. of the total cubic contents.

2. The cost, handling, freight and duty on the small and crooked ends which will be cut off before creosoting.

3. The cost, handling, freight and duty on the waste pieces occasioned by cutting piles to special lengths.

4. The cost, handling, freight and duty on piles which may be condemned by the Inspector as being unsound, small, or crooked.

5. One extra loading into the vessel, and one extra unloading from the vessel.

Purpose of creosoting.—The purpose of creosoting is to so fill up the cells of the wood that neither air, moisture, nor life can get inside. In order that this may obtain to the fullest possible extent, it is necessary that as little cutting as possible be done, and all cuts or broken surfaces be covered over with three or four coats of thick hot creosote, and where such surfaces are exposed above water they should be coated over once a year.

If two or three months elapse between the time of treatment and the using of the timber in actual construction, inject 1 to $1\frac{1}{2}$ lbs. of creosote extra per cubic foot of timber, to allow for evaporation.

Inspection.—Close inspection during the treatment by a reliable engineer, experienced in the work, is an absolute necessity, and honest contractors always prefer to have such a man at their works.

Economy in Construction.—The saving of labor for renewals and maintenance, and not the first cost, should be the ruling factor, and in a calculation for ultimate economy, it will be found a decided waste of money to expose untreated timber to the insatiable sea-worm. Public works and railroads exposing untreated timber to quick destruction by sea-worms, instead of defying them by using creosoted material, are neglecting an important economy ; in the case of public works spending the people's money for maintenance and renewals, and in the case of railroads spending money for mamtenance which should be paid in dividends to the stockholders. To secure good results, pay a sufficient price, contract with none but reliable persons, and keep an experienced person at the works during the preparation and treatment of the timber.

Price.—Creosoted piles, from 30 to 65 feet long, can be delivered in dock in almost any port in Canada, freight and duty paid, for from 30 to 40c. per lineal foot, according to lengths and sizes of butts and points ; square timber for from $35 to $45 per B.M., according to sizes, lengths and quantity of oil. Duty is 20 per cent.

· The following is a skeleton specification for the supply and creosoting of piles (short-leaf or Loblolly pine) with dead oil of coal-tar, which may prove useful as a general guide :

SPECIFICATION.

For piles of Virginia or North Carolina short-leaf or Loblolly pine, to be delivered, freight and duty (20 per cent.) paid, at ...., on or before the .... day of ...., 189..., as per accompanying bill.

The piles to be of the Virginia or North Carolina short-leaf pine, *Pinus mitis*, or Loblolly pine, *Pinus toedas*, sound, free from shakes, bad knots, or other imperfections that would reduce their strength. To be completed, barked and saw-butted, and so nearly straight that when the tape is stretched from the centres at the ends, it will not overhang the most crooked places more than one inch. They shall be not less than nine inches diameter at the small end, and not less than 16 inches diameter at the butt; all measurements after barking. To be of even taper and not button-headed, thus: a pile 16 inches at butt to be not less than 14 inches three feet down. The cubical contents of the round piles shall be determined by the rule used by the United States Government, as follows: Multiply the square of mean circumference by the length and divide by 12.5 ; if in inches divide again by 144.

Dead Oil of Coal-tar.—The creosote shall consist of dead oil of coal-tar heavier than water—not thin oil, lighter than water, thickened and weighted with coal-tar. The composition shall be as follows: At least two-thirds shall be obtained by distillation at a temperature exceeding 482° Fah., and the remainder at a temperature exceeding 392° Fah. Specific gravity at 60° Fah , between 1.035 and 1.065. Completely liquid at 100° Fah. To solidify between 40° and 45° Fah. Phenols between 5 per cent. and 10 per cent. Naphthaline not less than 50 per cent. Boiling point not below 410° Fah. The storage tank shall have no water on top of the creosote, nor no muddy nor objectionable matter at the bottom. When a sample for analysis is required, it shall be composed of one-half from the upper 12 inch layer and one-half from the lower 12-inch layer of the storage tank.

Treatment.—The piles shall be treated with not less than 16 pounds best dead oil of coal-tar per cubic foot. Sufficient percentage must be allowed for outside drainage when drawn out of the cylinders. To be subjected to heat by live and superheated steam, not over 250° Fah.; heat to be continued long enough to reach the centre of the timber. Vacuum to range between 22 and 25 inches. Heat above the condensing point shall be maintained in the cylinders during the vacuum and throughout the treatment, so that there shall be no condensation whatever at any stage of the process. Vacuum to be continued until the discharge from the pump shall have no odor or taste of turpentine, and kept up until the cylinders have been filled with oil. The oil in the gauge-tank connected with the force-pump to be measured at a temperature of about 125° Fah., which temperature should be maintained during the treatment. The vacuum in the cylinders must not be so suddenly produced, nor the temperature raised so high as to cause cracking or splitting of the timber. Cylinders to have proper outlets to free them from all gases before the pressure is applied. Pressure to be continued until the requisite quantity of oil has been injected into the timber. Water must not be admitted into the cylinders, nor allowed to accumulate in the upper part of the cylinders over the oil. Proper gauges to determine the amount of oil used, to be furnished. The piles shall be selected, and those having the same amount of seasoning, texture and density placed in the cylinder together, so that the penetration may be uniform. With every charge a test block shall be placed on the top of the load, which block will be afterwards split, and the penetration ascertained by the Inspector. The block should be $3\frac{1}{2}$ feet long by 12 inches in diameter, and as near as possible of the same degree of seasoning and texture as the piles forming the charge. The piles, creosote and treatment shall be subject to close inspection at the works and before shipment, and the whole of the materials and workmanship must be to the entire satisfaction of the Inspector. A chemical analysis of the oil must be furnished if requested.

PUBLICATIONS CONSULTED IN THE PREPARATION OF THIS PAPER.

" American Woods," by R. B. Hough, 1896

" Forest Wealth of Canada," by Geo. Johnson, 1895.

" Antiseptic Treatment of Timber," by J S. Boulton, 1884.

" Report of Committee of Am. Soc. C. E ," 1885.

I am also indebted to the following gentlemen for valuable information :

Geo. S. Valentine, manager, Eppinger & Russell, Creosoting Works, New York.

B. Comer, superintendent, Lehigh Valley Creosoting Co., New York.

Dr. Henry Froehling, analytical chemist, Richmond, Va.

A. S. Martin, manager, Old Dominion Creosoting Works, Norfolk, Va.

Edmund Christian, engineer and general manager, Norfolk Creosoting Co , Norfolk, Va.

H. S. Haines, of the Plant R. R. and Steamship System.

B. T. Burchardi, chief engineer and general manager, Fernandina Oil and Creosote Works, Fernandina, Fla.

H. J. Mackenzie, C.E., Creosoting Inspector, Moncton, N B.

CPSIA information can be obtained
at www.ICGtesting.com
Printed in the USA
BVHW04s1224011018
528938BV00023B/1218/P